TOP OF THE OCEAN FOOD WEB:
Orcas & Great White Sharks

WRITTEN BY MATT REHER

ILLUSTRATED BY LI LIU

The ocean is full of animals that have to eat. Some ocean animals eat plants. Some eat lots of tiny fish. Some eat bigger fish. This is a food web that shows who eats what.

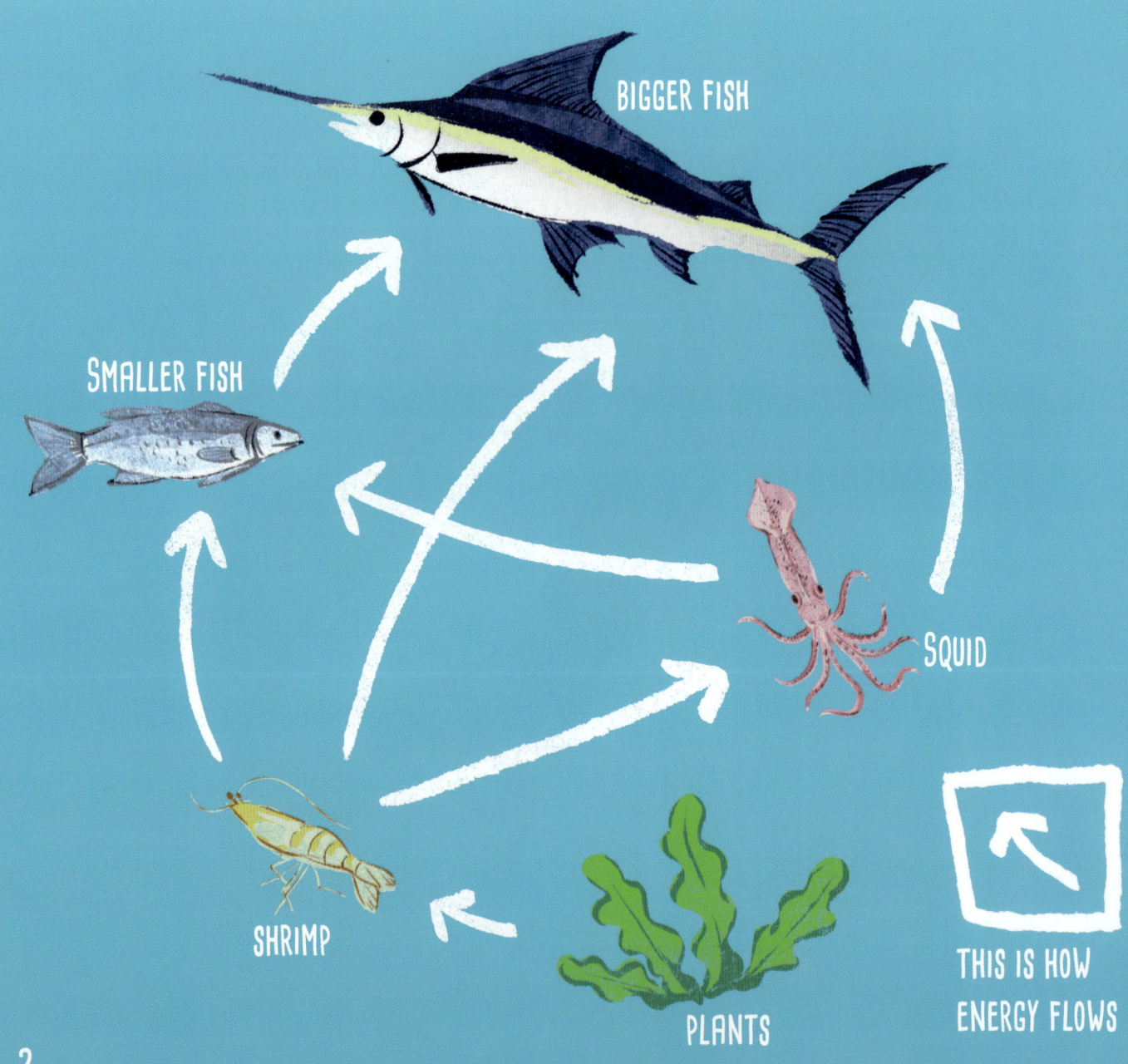

Orcas and great white sharks are the two biggest and baddest eaters in the whole ocean. They both can eat almost any animal in the ocean. But which one is at the top of the ocean food web?

WHICH ANIMAL IS BIGGER?

32 Feet

Male orcas can grow up to 32 feet long. That's as long as a fire truck.

WHO WINS?

ORCAS-1 SHARKS-0

ORCAS ARE BIGGER!

Great white sharks can grow up to 20 feet long. That's the size of a pickup truck.

WHO HAS THE SHARPEST TEETH?

Orcas have 40-50 large teeth. They are shaped like cones. They are not very sharp. If an orca loses a tooth, it will not grow back.

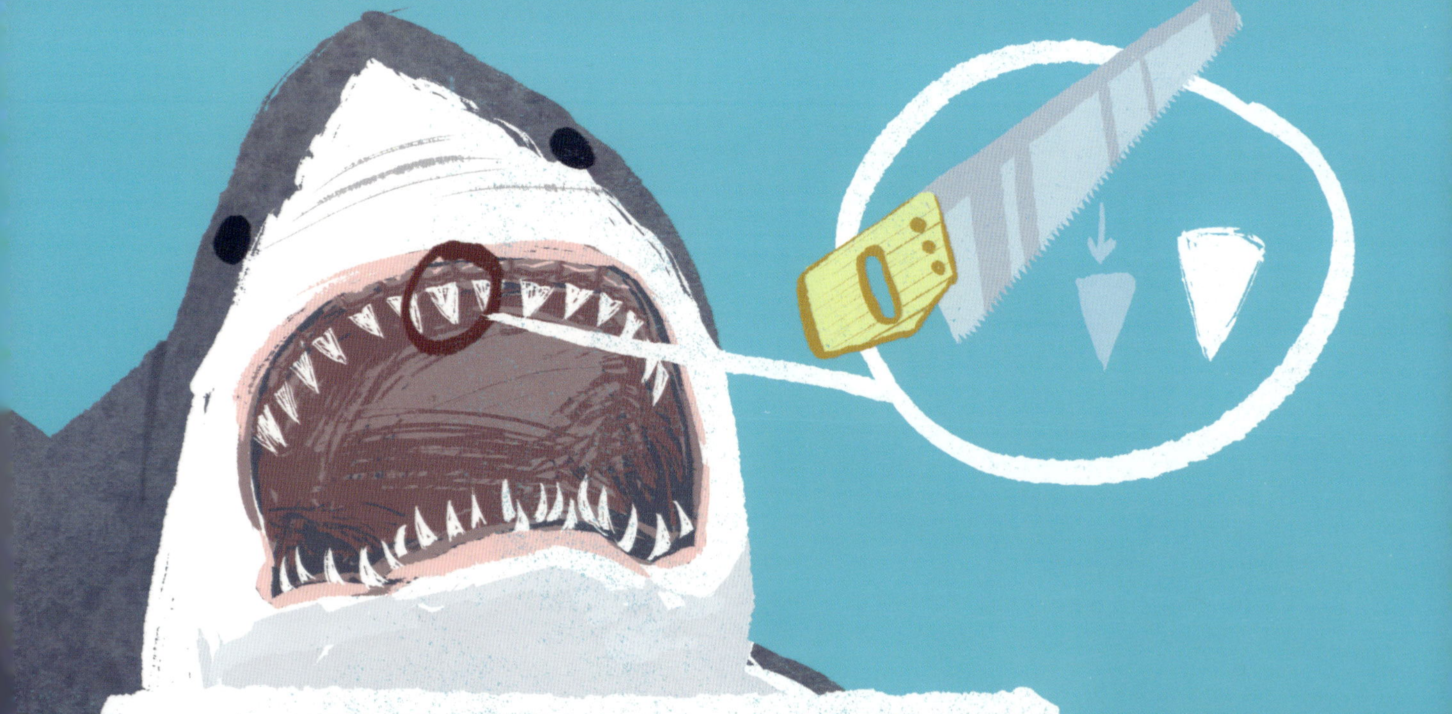

Great white sharks have about 3,000 teeth. These teeth are very sharp. They have many rows of teeth. If a tooth comes out, another tooth will take its place.

A GREAT WHITE SHARK's BITE REALLY HURTS!

WHO WINS?

ORCAS - 1   SHARKS - 1

WHICH ANIMAL IS FASTER?

25 MPH

Sharks are slow swimmers most of the time. When sharks hunt, they can swim very fast.

Just like the great white shark, orcas only swim fast when they are trying to catch food. An orca can swim faster than a great white shark.

WHO WINS?

25 MPH   30 MPH

30 | 25 MPH
ORCAS-2 SHARKS-1

ORCAS ARE MUCH FASTER!

WHO IS THE BETTER HUNTER?

Orcas are very smart hunters. They hunt in groups called pods. They work together to trap seals. Sometimes they even make large waves to knock seals into the water.

A SMART POD OF **ORCAS** ALMOST ALWAYS GETS THE ANIMALS IT HUNTS!

A SHARK CAN SMELL A FISH FROM ALMOST 100 FT AWAY. THAT'S THE LENGTH OF A BASKETBALL COURT!

WHO WINS?
POD — ORCAS-3
ALONE — SHARKS-1

Sharks hunt alone. Sharks can't see very far, so they need their noses to catch fish. They can smell a fish from far away. Then they swim fast to catch it with their sharp teeth.

Most people think that the great white shark is the biggest and baddest animal in the ocean. Orcas are bigger. They are faster. They are better hunters. Does that make orcas the top of the ocean food web?

Here is the test.
Does one of these
big, bad animals
eat the other?

Yes! When orcas in a pod see a great white shark, they swim under it. Then, they move their tails back and forth. Orcas have strong tails.

This pushes the water up. It also pushes the shark up to the top of the water.

Then the orca grabs the shark and flips him upside down. When sharks are upside down, they are stuck. They can't move. They can't fight back, and they can't breathe.

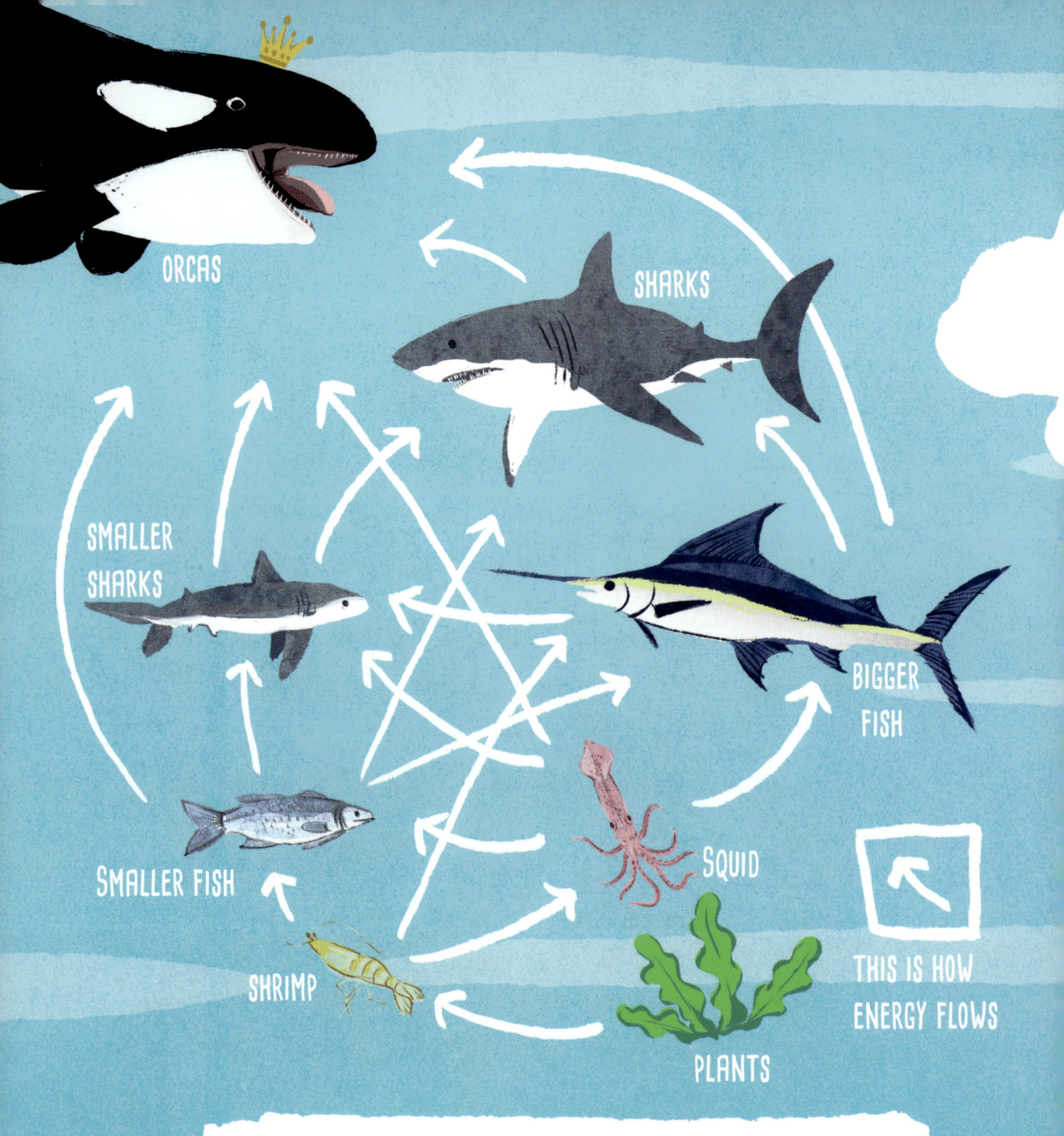

The orca will eat the shark. The orca is the top eater of the ocean food web.

FUN FACTS ABOUT SHARKS AND ORCAS

Many sharks are scared of orcas. This is why many sharks have started to stay away from the parts of the ocean where orcas live and hunt.

Some sharks hunt near beaches where children play. To make beaches safer, some beaches play orca sounds in the water. This noise should keep the sharks away.

ORCA MAP

orca habitat

SHARK MAP

shark habitat